My First Science Books

A CRABTREE SEEDLINGS BOOK

What Is A SHARK?

Dominic Carter

A Crabtree Seedlings Book

Crabtree Publishing

crabtreebooks.com

Sharks live in the ocean.

Their **gills** allow them to breathe under water.

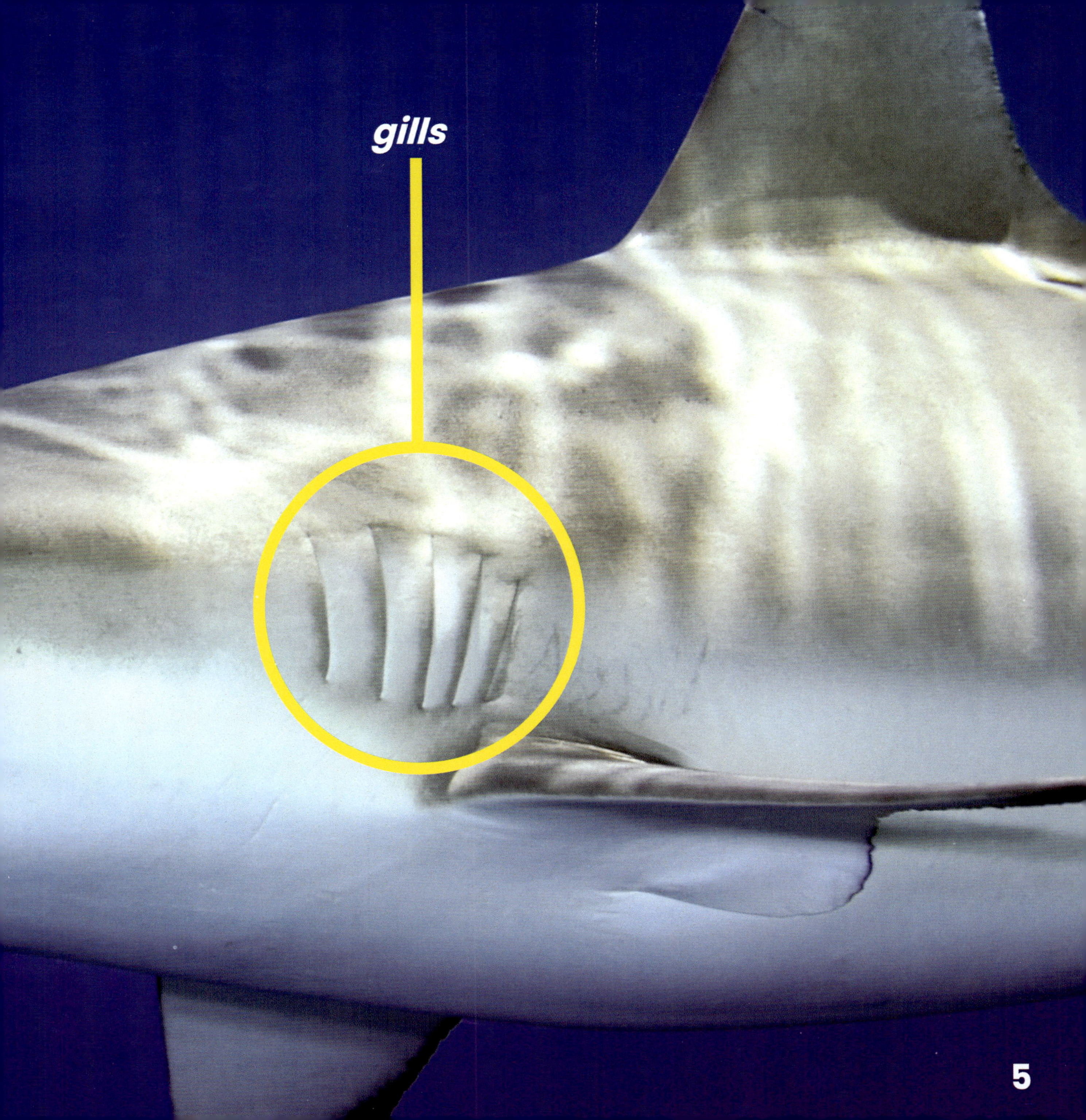
gills

Sharks have strong, stiff **fins** to help them swim fast.

first dorsal fin
pectoral fin

A shark's skin is covered with small, toothlike **scales**.

Most sharks have rows of sharp teeth.

A shark is a **fish.**

Fish have things in common.

All fish live in water. They use their fins to swim. All fish have gills. Most fish have scales.

Some sharks are born **live**.
Others hatch from eggs.

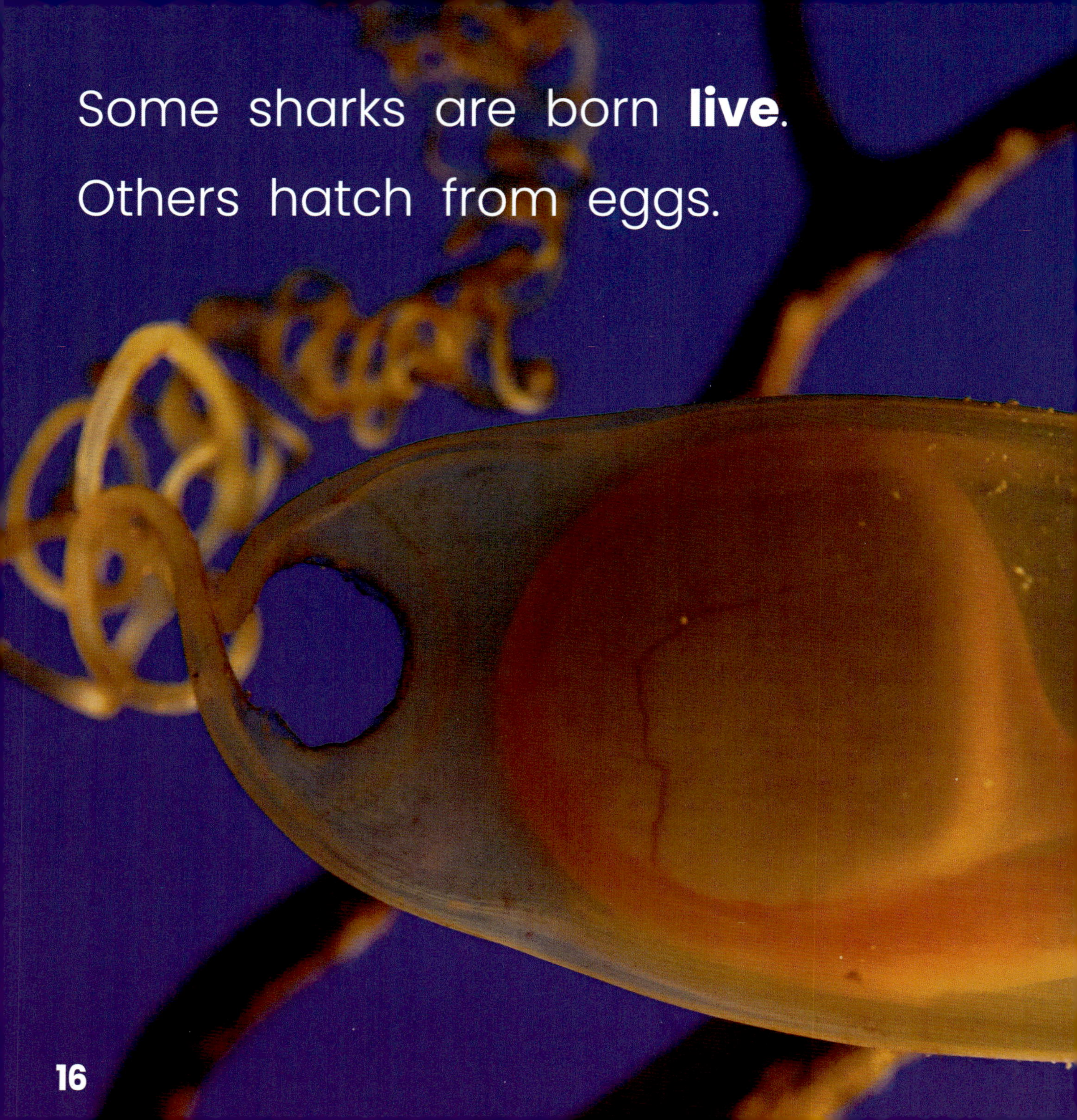

shark egg

The whale shark is the biggest of all sharks.

Some whale sharks are as big as a school bus!

Mako sharks are the fastest of all sharks.

Mako sharks
can also jump really
high—as high as
20 feet
(6 meters)!

Glossary

fins (FINZ): Fins are the body parts that fish use to help them swim and turn.

fish (FISH): A fish is a cold-blooded animal that lives in water and has scales, fins, and gills.

gills (GILZ): Gills are the body parts that fish use to breathe.

live (LYVE): Live babies grow inside their mother and are born from their mother—not hatched from an egg.

scales (SKAYLZ): Scales are small pieces of hard skin that cover the bodies of fish and reptiles.

Index

School-to-Home Support for Caregivers and Teachers

Crabtree Seedlings books help children grow by letting them practice reading. Here are a few guiding questions to help the reader with building his or her comprehension skills. Possible answers are included.

Before Reading

- **What do I think this book is about?** I think this book is about sharks. It might tell us about different kinds of sharks.
- **What do I want to learn about this topic?** I want to learn about shark body parts.

During Reading

- **I wonder why...** I wonder why mako sharks jump in the air.
- **What have I learned so far?** I have learned that sharks have gills, fins, scales, and sharp teeth.

After Reading

- **What details did I learn about this topic?** I learned that the whale shark is the biggest type of shark. Some are as big as a school bus!
- **Read the book again and look for the vocabulary words.** I see the word ***fish*** on page 12 and the word ***live*** on page 16. The other vocabulary words are found on pages 22 and 23.

Crabtree Publishing

crabtreebooks.com 800-387-7650

In Canada: We acknowledge the financial support of the Government of Canada through the Canada Book Fund for our publishing activities.

Published in Canada
Crabtree Publishing
616 Welland Avenue
St. Catharines, Ontario
L2M 5V6

Published in the United States
Crabtree Publishing
347 Fifth Avenue
Suite 1402-145
New York, NY 10016

Written by: Dominic Carter
Production manager: Candice Campbell
Photo Credits: Cover: boy © Gelpi, shark © Matt9122, thought bubble art © doddis77; page 4-5 © Rich Carey; page 6-7 © cbpix; page 8-9 © A Cotton Photo; page 10-11 © Matt9122; page 12-13 © Greg Amptman; page 14-15 © Gelpi, thought bubble art © doddis77; page 16-17 © magnusdeepbelow; page 18-19 © Alessandro De Maddalena; All photos from Shutterstock.com except inset photo page 11 © Pascal Deynat/Odontobase; https://creativecommons.org/licenses/by-sa/3.0/deed.en ; page 20-21 © Sander van der Wel from Netherlands https://creativecommons.org/licenses/by-sa/2.0/deed.en

Hardcover 978-1-4271-3030-3
Paperback 978-1-4271-3041-9

Library and Archives Canada
Cataloguing in Publication
Available at the Library and Archives Canada

Library of Congress
Cataloging-in-Publication Data
Available at the Library of Congress

Printed in the U.S.A./CP062026